The Learning Works

Middle School Science Challenge

Written by Kathy Harbaugh

The Learning Works

Dedicated to
Bob, Erin, and Danica
———o———
Special thanks to
Pat Bixler and Janey Cohen

Editorial production & page design by
Clark Editorial & Design

Illustrations by *Marcy Ramsey*

Copyright © 1996
THE LEARNING WORKS, INC.
P.O. Box 6187
Santa Barbara, California 93160
All rights reserved.
Printed in the United States of America.

ISBN: 0-88160-273-6

Introduction

Middle School Science Challenge is a collection of fun and easy-to-use questions, activities, and experiments designed especially for the middle school classroom. *Middle School Science Challenge* is divided into six high-interest areas:

- Fascinating Facts
- Famous People
- Name That Thing

- Science Explorations
- Field and Classroom Experiments
- Discussion Starters

Middle School Science Challenge contains instant activities that can be used in a wide variety of ways, such as:

- class openers
- "problem-of-the-day" challenges
- supplements to class topics
- learning center activities

- extra credit
- homework assignments
- motivation for research and discovery
- bonus points on unit exams

The questions can be reproduced or read aloud as brain teasers to get students thinking, while the hands-on activities and experiments make ideal cooperative learning exercises.

Using the Answer Key

No collection of questions would be complete without a collection of answers. In this book, the answer key appears at the back of the book, on pages 127–136. To make needed answers easier to find, they have been listed by page number and keyed by letter to a particular position on the page. Thus, answer **a** is for the question on the upper left side of the page, answer **b** is for the question on the upper right side of the page, answer **c** is for the question on the lower left of the page, and answer **d** is for the question on the lower right side of the page. (See diagram.)

a	b
c	d

Middle School Science Challenge is a fun-filled book you and your students will enjoy all year long!

Contents

Contents
(continued)

Graph of Scientific Areas

Page	10	11	12	13	14	15	16	17	18	19	20	21	22	23	24	25	26	27	28	30
Earth	•		•				•				•					•			•	
Life		•			•			•		•		•		•			•			
Physical				•		•			•				•		•			•		•

Page	31	32	33	34	35	36	37	38	39	40	41	42	43	44	45	46	47	48	50	51
Earth		•				•			•			•								
Life	•			•		•					•		•		•	•		•	•	
Physical			•	•				•		•				•			•			•

Page	52	53	54	55	56	57	58	59	60	61	62	63	64	65	66	67	68	70	71	72
Earth	•		•				•			•				•		•				
Life		•			•			•			•		•				•	•	•	
Physical				•		•			•			•			•					•

Graph of Scientific Areas
(continued)

Page	73	74	75	76	77	78	79	80	81	82	83	84	85	86	87	88	89	90	91	92
Earth				•	•	•	•					•	•			•	•			
Life		•	•					•	•									•	•	
Physical	•									•	•			•	•					•

Page	94	95	96	97	98	99	100	101	102	103	104	105	106	107	108	109	110	111	112	113
Earth									•	•					•	•				
Life	•	•					•	•					•	•			•	•		
Physical			•	•	•	•					•	•							•	•

Page	114	115	116	118	119	120	121	122	123	124	125	126
Earth	•	•	•	•		•			•	•	•	•
Life				•	•	•	•	•				•
Physical				•				•				

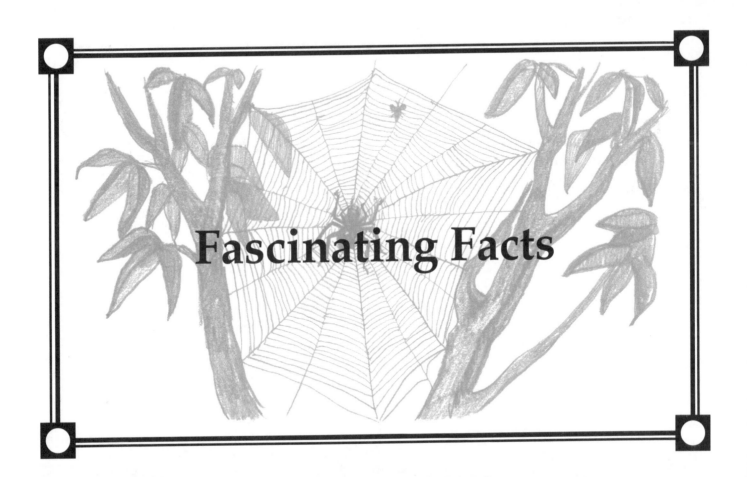

Fascinating Facts

Which planet rotates so fast that its day is only 10 hours long?

What composes the colored rings of Saturn?

What makes up Uranus's soft, blue-green color?

What percentage of sunlight received on Earth is received on Neptune?

What causes some organisms
to glow in the dark?

How many cells make up
the human body?

What very thin thread produced
by an animal can stretch by a third
and would have to be 50 miles
long to break under
its own weight?

Which male marine animal
incubates and gives birth
to live offspring?

Middle School Science Challenge
© The Learning Works, Inc.

What was first discovered about the planet Uranus in 1977?

Which continent has no land areas below sea level?

How much of the earth's land surface is located north of the equator?

What is the world's record high temperature and where was it recorded?

What is the shape
of all snowflakes?

What were the first five metallic
elements that the ancient
Egyptians discovered?

Why is a red laser beam red?

Why is frozen carbon dioxide
called "dry ice"?

13

What was the first
form of life on Earth?

What is required for goldfish
to retain their bright color?

How many toes
did the first horse have?

What animal
has transparent blood?

In which phase
is water the densest?

Which chemical, when dropped
into water, will cause
an explosion?

Why are dams made thicker
at the bottom than at the top?

What did Archimedes discover
when he stepped into his bath?

15

Where and in what form
is most of the earth's fresh water?

If all of the ice of Antarctica
melted, how much of Earth's land
would be underwater?

What is the name of the stream
of water that warms the shores
of western Europe, and where
does it begin?

Which volcanic island disappeared
from the map in 1883 when it
exploded with such force that
it became a 900-foot chasm
underwater?

What nonsighted fish uses its body to generate 400 volts of electricity for defense and catching its prey?

Which insect can lift something fifty times its own body weight?

What animal can run at speeds of more than 60 miles per hour and is considered the fastest animal on Earth?

Which marine mammal's heart beats approximately seven times per minute, compared with the human heartbeat of about 70 times per minute?

What part of a gas flame is invisible, giving off no light at all?

At high altitudes, why does it take hours to boil food?

Where could you see a rainbow form a complete circle?

What is the most common metallic element on Earth, although it only exists in chemical combination with other elements?

Humans with which color hair have the most hairs on their heads?

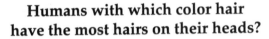

Cells are constantly being rubbed off the human skin. How long does it take for a completely new outer layer of skin to form?

What fish can survive out of water for three years?

Rattlesnakes have no sense of hearing. How do they sense an approaching enemy or prey?

Middle School Science Challenge
© The Learning Works, Inc.

How large, relative to Earth, is the Great Red Spot on Jupiter?

How have scientists documented 160,000 years of the climate on Earth?

What would happen to the climate of the Amazon basin if all of the rain forests were converted to pasture land?

Which mineral is believed to be the oldest on Earth?

Why do people often perspire
when they eat spicy foods?

What is the hardest substance
in the human body?

Which explosive is sometimes
used to treat a heart condition?

What name is given to the phase
of sleep during which the body
enters a temporary state
of benign paralysis?

Middle School Science Challenge
© The Learning Works, Inc.

Which element has the shortest liquid range, the difference between its melting point and its boiling point being 4.6°F?

Where would the highest known temperature of 540 to 720 million degrees Fahrenheit be found on Earth?

What was used to produce the highest pitch sound?

A column of air one inch square and 600 miles high weighs approximately 15 pounds. What scientific term is used to describe this?

Where would you find
Hydrophis belcheri, the most
venomous snake in the world?

Which animal can eat enough
leaves in its first 56 days of life to
equal 86,000 times its birth weight,
making it the most prodigious
eater known?

In proportion to their size, what
animals are the strongest?

What is the largest cell
in the human body?

What phenomenon, also the basis of police radar, makes light or sound waves increase in frequency if the source is moving toward the observer and decrease if the source is moving away?

What type of simple machine did the Egyptians use to build the pyramids?

Which colorless, odorless, tasteless gaseous element, rare in the earth's atmosphere, is believed to have the potential to cause cancer?

One ounce of which metal can be drawn (stretched out) to a length of 43 miles?

What is the name of the display of colored lights in the sky near the North Pole that is caused by the interaction of the upper atmosphere with particles from the sun?

What could alter *Voyager* 2's course into space, farther and farther away from our solar system?

Which planet is still made up of the original gas and dust that congealed to form the sun and planets?

Which country experiences virtually no tide?

Middle School Science Challenge
© The Learning Works, Inc.

What organ in the human body is 2% of the body's weight but requires 25% of the oxygen used by the body?

Which bird is the only one capable of flying backwards?

In which part of the human anatomy are one-quarter of the bones located?

What other animal loses the hair on its head the way humans do?

What popular beverage
did Joseph Priestley invent
while experimenting with air?

How can an airplane reduce
the amount of fuel it uses?

What liquid has no measurable
viscosity, meaning it has the
ability to flow with complete
absence of friction and remain
in a liquid state at absolute zero?

What is the name of the principle
that enables scientists to accelerate
objects so fast that they increase
in mass?

Where are the highest tides, ranging up to 47.5 feet?

What natural phenomenon can travel across the ocean at a speed of up to 490 miles per hour?

Where is the wettest place in the world, with an average annual rainfall of 463.4 inches?

What are the oldest known rocks on Earth?

Famous People

What scientist discovered radium in the early twentieth century and was the only person ever to win two Nobel Prizes in two different sciences—chemistry and physics?

Who first demonstrated, with the use of prisms, that sunlight is a mixture of light of all colors?

Who published the first periodic table of elements, which appeared in 1869?

Which English physicist first demonstrated that heat is a form of energy?

James Watson and Francis Crick
built the first ladder-like model
of DNA in 1953 based on the x-rays
of which scientist?

Which married team of scientists
discovered the process of
metabolism, and shared
the Nobel Prize in physiology
and medicine in 1947?

Who was the world's first
test tube baby?

In the 400s B.C., who taught that
diseases have natural causes and
that the body can repair itself?

Who discovered the Andromeda Galaxy using the 100-inch, Mount Wilson telescope?

Which African-American scientist greatly impressed Thomas Jefferson by making the astronomical and tide calculations as well as weather predictions for a yearly almanac from 1791 to 1796?

Which scientist is considered to be the founder of modern geology because he argued that the earth's history could be read in rocks?

Which NASA researcher warned a Senate committee in 1988 that increased levels of carbon dioxide were causing global warming (the greenhouse effect) and that melted polar ice could lead to great flooding?

Who hypothesized the law
that explains that equal volumes
of all gases contain equal numbers
of molecules at the same
temperature and pressure?

In the 1700s, who invented
the modern version of the steam
engine and had a unit of power
named after him?

Which two men achieved
the first sustained flight of a
heavier-than-air machine in 1903?

Which inventor of optical
processing procedures became
the first Hispanic-American
female astronaut in 1990?

Middle School Science Challenge
© The Learning Works, Inc.

In the 200s B.C.,
who discovered the laws
of the lever and the pulley?

Which German physicist advanced
the theory that energy is given off
in a stream of separate units
called quanta?

Which scientist was the first
to make an X-ray of vitamin B-12
and earned the Nobel Prize in
chemistry in 1964?

Who revolutionized scientific
thinking about space and time
with his Theory of Relativity?

Which surgeon general, appointed in 1990, was a strong advocate for women and children, particularly those afflicted with AIDS and teenagers with drinking problems?

Who discovered penicillin, the first antibiotic?

Which scientist used peanuts and sweet potatoes to improve the land and diversify the economy of the southern states?

Which marine biologist and gifted writer publicized the dangers of the pesticide DDT in 1962?

Who became the first African-American woman to travel in space in September of 1992 aboard space shuttle *Endeavour*?

In the first century A.D., who proposed the geocentric theory, which stated that the earth is the center of the universe?

In 1543, who published a book suggesting that the sun was the center of the solar system, the heliocentric theory?

Who used intricate calculations to show that the heliocentric theory of the solar system would work if the planets orbited the sun in elliptical orbits rather than circular ones?

Who produced the first effective polio vaccine in 1953?

In 1910, which American biologist showed that genes are the units of heredity and that they are arranged in an exact order along chromosomes?

Which doctor opened a free medical clinic in Los Angeles for Hispanic Americans and established scholarships for Hispanic Americans to study medicine?

In the mid-nineteenth century, who discovered that certain kinds of microscopic organisms cause disease?

Middle School Science Challenge
© The Learning Works, Inc.

In 1852, who had the idea to combine dots and dashes to represent the alphabet with his telegraph?

Which crew member aboard the 1986 flight of the space shuttle *Columbia* conducted experiments as a physicist and was the first Hispanic American to travel in space?

In 1960, which scientist created the first laser beam?

Which scientist used a mixture of ice and salt to achieve the lowest temperature possible, calling it zero degrees F?

Which crew member aboard the *Vostock 6*, launched in 1963 by the Soviet Union, was the first woman in space?

Who were the first two Americans to set foot on the moon?

Who first charted the Gulf Stream in the eighteenth century?

During the space shuttle *Challenger* mission launched in August of 1983, who became the first African American to fly in space?

39

The unit of electrical resistance is named for which scientist?

Which nineteenth-century English physicist calculated that when all the heat had been taken out of a source material, the temperature would be –273°C, absolute zero, which led to the development of the temperature scale which bears his name?

Which two people founded Apple Computer by first building prototypes in a garage?

In 1937, which American pilot attempted to fly around the world in her plane, the *Electra*?

In 1984, which two scientists independently identified the agent responsible for HIV?

Eugenie Clark is a leading authority on shark behavior. What name is given to Clark and other scientists who study fish?

Who is responsible for discovering how to preserve blood plasma in blood banks so that it is ready for transfusions when needed?

During the 1500s, which well-known artist made great contributions to the sciences of anatomy and botany as well as astronomy and geology?

41

Based on evidence that the universe is still expanding, which two scientists in 1946 proposed the Big Bang theory of the creation of the universe?

In 1967, which two astronomers discovered pulsars (pulsating neutron stars), previously known as "little green men" because of the extraterrestrial signals the astronomers received from them?

In the 1500s and 1600s, who was the most influential person in setting the Scientific Revolution in motion, pulling modern science out of ancient natural philosophy?

Who is known as "America's first woman astronomer" and what did she discover in 1847?

Famous People

Which well-known dermatologist
is responsible for great
contributions to the scientific
treatment of syphilis and leprosy?

In 1602, who became the founder
of modern physiology with his
discovery that blood circulates?

Jane Goodall is well known
for her behavioral studies
of which animals?

apes.

During the late 1800s and early
1900s, which botanist had 15
species of plants and mosses
named after her and founded
the Wild Flower Preservation
Society of America?

43

Middle School Science Challenge
© The Learning Works, Inc.

In the mid-1800s, who worked out the equations which explain the laws of electricity and magnetism?

Which physicist co-discovered nuclear fission but did not receive a Nobel Prize as did her colleague?

What inventor earned more than 35 patents in his lifetime and is best known for his many inventions having to do with the electrical control and distribution of railways?

In 1935, what scientist and her husband brought a third Nobel Prize to her family for the discovery of artificial radioisotopes?

Which Cornell professor
illustrated 600 insects on wood
engravings and, in 1911, published
the *Handbook on Nature Study*
(known as the "Nature Bible")?

Which two physicians first
isolated the hormone insulin,
used in treating diabetes?

In 1893, who became the first
doctor to operate on the heart
and was instrumental in
the advancement of African-
American physicians?

In 1959, who discovered skull
fragments from an early ancestor
of modern humans, believed to be
almost two million years old?

Middle School Science Challenge
© The Learning Works, Inc.

Between 1904 and 1915, who continued to spread typhoid in epidemic proportions throughout New York City, but was immune to it herself?

Which African-American scientist received recognition in 1915 for his pioneer investigations into the mysteries of egg fertilization and the study of the cell?

Which marine biologist and former governor of Washington became the first woman to chair the Atomic Energy Commission in 1973 because of her concern that more knowledgeable scientists be represented in public office?

What Russian physiologist researched reflex conditioning that involved the pairing of stimuli to produce a reflex?

Which chemist and meteorologist computed a radioactive fallout pattern formula still used at every naval station to determine the safest evacuation route for that area?

A unit of electrical energy is called a volt in honor of which scientist?

In 1958, who assembled the first microchip, permitting the development of tiny electronic circuits capable of carrying out complex tasks?

Who used a burning candle to prove the Law of Conservation of Matter in 1777?

Who wrote the first scientific text on human anatomy, *On the Fabric of the Human Body*, published in 1543?

Who wrote *The Origin of Species*, proposing the theory of evolution?

During the early 1900s, which nurse launched her lifelong campaign for birth control?

Who first saw cells in the microscope in the mid-1600s?

Name That Thing

This body system regulates the passage of oxygen and nutrients to the cells and takes wastes away from them.

the circulatory system

A written recording of the electrical activity of the brain is produced by this.

Oxygen is carried in the blood by this complex protein.

blood cells

Many living organisms follow this activity cycle, which lasts 24 hours.

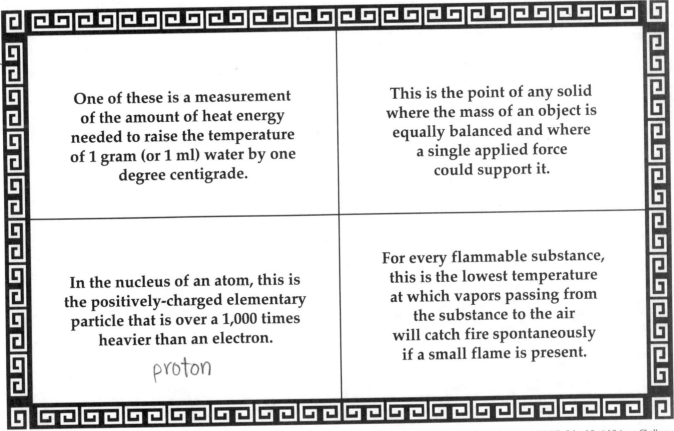

One of these is a measurement of the amount of heat energy needed to raise the temperature of 1 gram (or 1 ml) water by one degree centigrade.

This is the point of any solid where the mass of an object is equally balanced and where a single applied force could support it.

In the nucleus of an atom, this is the positively-charged elementary particle that is over a 1,000 times heavier than an electron.

proton

For every flammable substance, this is the lowest temperature at which vapors passing from the substance to the air will catch fire spontaneously if a small flame is present.

Middle School Social Science Challenge
© The Learning Works, Inc.

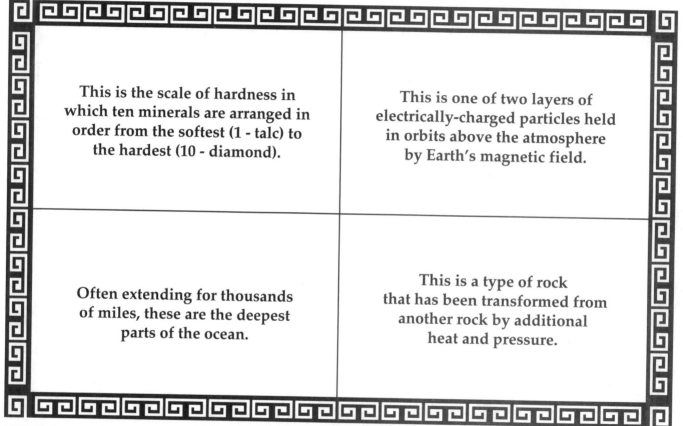

This is the scale of hardness in which ten minerals are arranged in order from the softest (1 - talc) to the hardest (10 - diamond).

This is one of two layers of electrically-charged particles held in orbits above the atmosphere by Earth's magnetic field.

Often extending for thousands of miles, these are the deepest parts of the ocean.

This is a type of rock that has been transformed from another rock by additional heat and pressure.

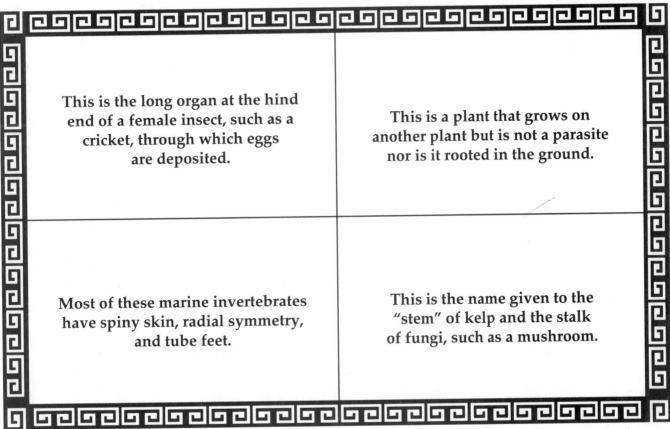

This is the long organ at the hind end of a female insect, such as a cricket, through which eggs are deposited.

This is a plant that grows on another plant but is not a parasite nor is it rooted in the ground.

Most of these marine invertebrates have spiny skin, radial symmetry, and tube feet.

This is the name given to the "stem" of kelp and the stalk of fungi, such as a mushroom.

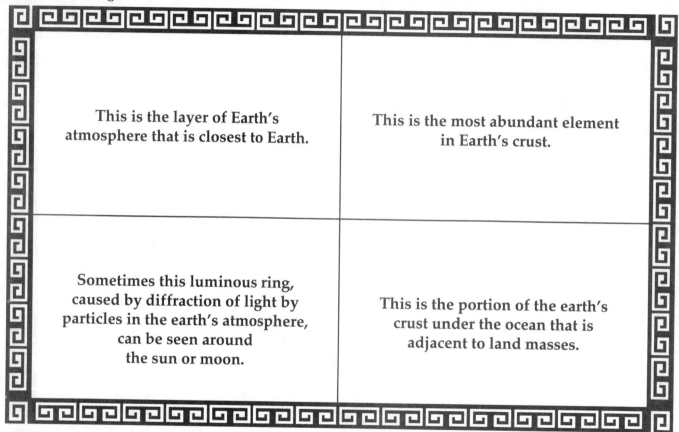

This is the layer of Earth's atmosphere that is closest to Earth.

This is the most abundant element in Earth's crust.

Sometimes this luminous ring, caused by diffraction of light by particles in the earth's atmosphere, can be seen around the sun or moon.

This is the portion of the earth's crust under the ocean that is adjacent to land masses.

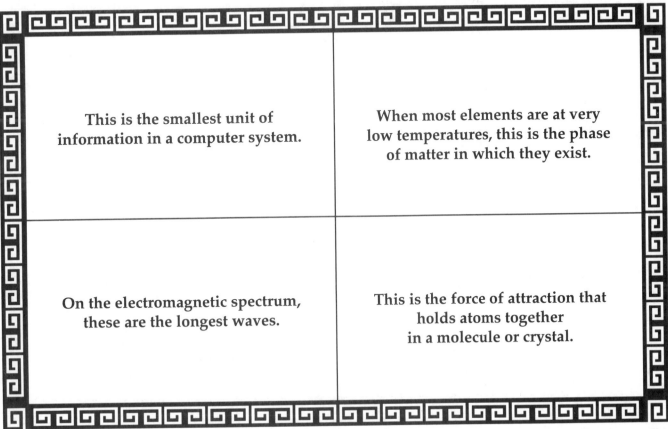

This is the smallest unit of information in a computer system.

When most elements are at very low temperatures, this is the phase of matter in which they exist.

On the electromagnetic spectrum, these are the longest waves.

This is the force of attraction that holds atoms together in a molecule or crystal.

These organisms are really symbiotic associations between fungi and algae.

Porifera is the phylum name for these simple marine organisms.

This describes an animal's automatic and innate response to a particular stimulus.

In vertebrates, this is the body part responsible for absorbing water and minerals from indigestible food and forming the solid waste which the animal expels.

Pivoting on a fulcrum,
this simple machine generally
is a rigid bar.

One element, plus oxygen
(a binary compound), forms this
kind of compound.

This is an instrument
for measuring the density
or relative density of liquids.

This is the angle between
a wave front and a surface
that it strikes.

Middle School Social Science Challenge
© The Learning Works, Inc.

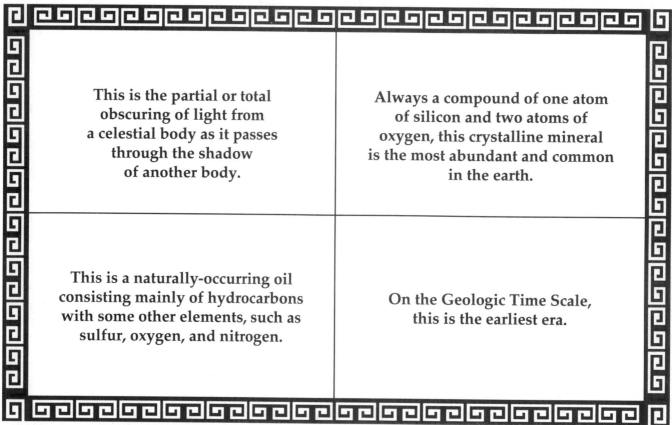

This is the partial or total obscuring of light from a celestial body as it passes through the shadow of another body.

Always a compound of one atom of silicon and two atoms of oxygen, this crystalline mineral is the most abundant and common in the earth.

This is a naturally-occurring oil consisting mainly of hydrocarbons with some other elements, such as sulfur, oxygen, and nitrogen.

On the Geologic Time Scale, this is the earliest era.

This phylum of invertebrate animals is comprised of more than one million species, making it the largest and most diverse phylum in the animal kingdom.

Some Protozoans, such as *Amoeba*, use this outgrowth of the cell which serves as a feeding and locomotion organ.

This type of organism always makes up the first level (base) of a food pyramid and always begins a food chain.

Mammals and reptiles have these tiny sacs in their lungs where respiratory gases are exchanged.

Middle School Social Science Challenge
© The Learning Works, Inc.

This is a material made of two or more metallic elements.

These objects in space appear starlike on optical photos, but have large redshifts, quite unlike stars.

This force along the surface of a liquid can cause that liquid to form droplets if the force is great enough for that particular liquid.

An object has this type of energy based on its position or structure rather than its motion.

Three-quarters of the world's earthquakes and considerable volcanic activity occur in or near the Pacific Ocean. This is the name scientists have given to this region.

This is a body of water formed when a river enters the ocean, mixing the fresh water with the salt water.

These objects from space burn up in Earth's atmosphere and are often called "shooting stars."

When a glacier retreats, this ridge of rock fragments remains.

Middle School Social Science Challenge
© The Learning Works, Inc.

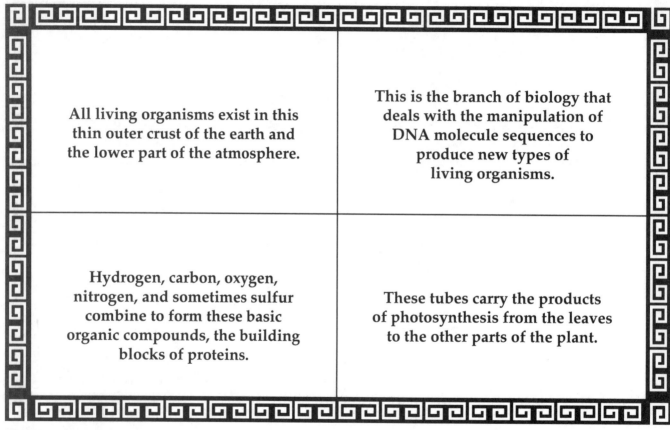

All living organisms exist in this thin outer crust of the earth and the lower part of the atmosphere.

This is the branch of biology that deals with the manipulation of DNA molecule sequences to produce new types of living organisms.

Hydrogen, carbon, oxygen, nitrogen, and sometimes sulfur combine to form these basic organic compounds, the building blocks of proteins.

These tubes carry the products of photosynthesis from the leaves to the other parts of the plant.

A material changes from the gas
phase to the liquid phase
at this temperature.

A great deal of energy can be
released by this nuclear reaction
when a large nucleus is split into
two or more smaller nuclei.

This is the name given to the
number of protons always found
in an atom of a particular element.

This is the process
of passing an electrical current
through a material to produce
a chemical breakdown reaction.

Middle School Social Science Challenge
© The Learning Works, Inc.

This is the rate at which
an organism at rest
expends energy.

These most primitive life forms
depend upon other living cells
to reproduce and grow.

Humans have a total of 46 of these
chemical sets of instructions
in the nuclei of their cells.

The pollen grows
on this organ of a flower.

Rocks consist of these inorganic substances that have regular internal structures and definite chemical compositions.

This is the high point of a continent where the waters flow in opposite directions to different oceans or bodies of water.

These narrow tubes of high-altitude air move more rapidly than the surrounding air and change position within a certain range.

Earth's crust, down to about 30 miles, is broken up into these dozen or more pieces.

Middle School Social Science Challenge
© The Learning Works, Inc.

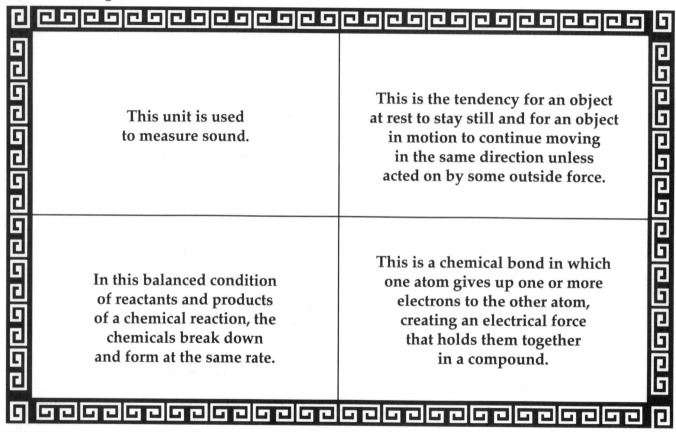

This unit is used
to measure sound.

This is the tendency for an object
at rest to stay still and for an object
in motion to continue moving
in the same direction unless
acted on by some outside force.

In this balanced condition
of reactants and products
of a chemical reaction, the
chemicals break down
and form at the same rate.

This is a chemical bond in which
one atom gives up one or more
electrons to the other atom,
creating an electrical force
that holds them together
in a compound.

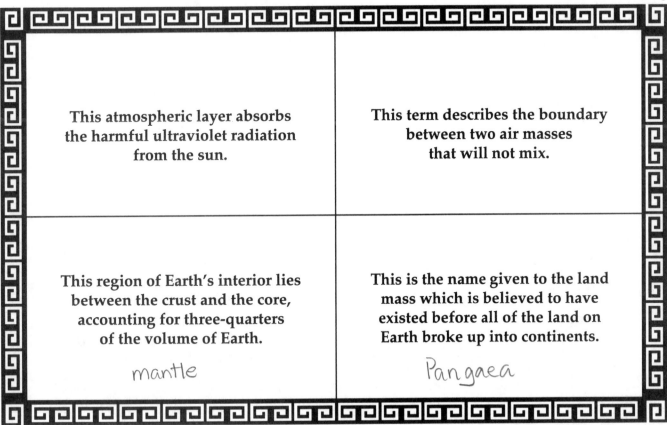

This atmospheric layer absorbs the harmful ultraviolet radiation from the sun.

This term describes the boundary between two air masses that will not mix.

This region of Earth's interior lies between the crust and the core, accounting for three-quarters of the volume of Earth.

mantle

This is the name given to the land mass which is believed to have existed before all of the land on Earth broke up into continents.

Pangaea

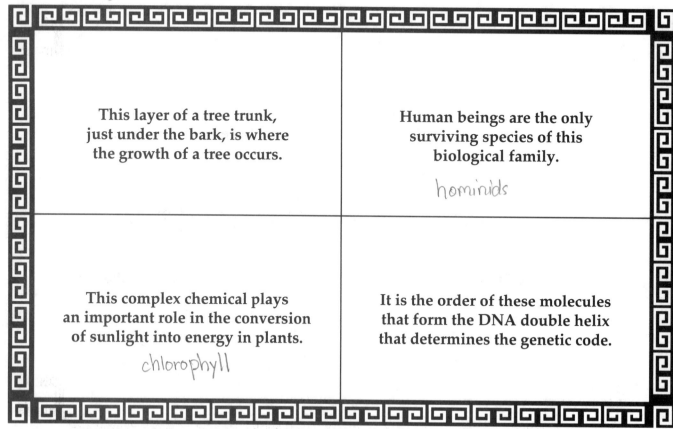

This layer of a tree trunk, just under the bark, is where the growth of a tree occurs.

Human beings are the only surviving species of this biological family.

hominids

This complex chemical plays an important role in the conversion of sunlight into energy in plants.

chlorophyll

It is the order of these molecules that form the DNA double helix that determines the genetic code.

Science Explorations

Cell Salad

Cells are the building blocks of all living organisms. The more complex animals may have hundreds of different types of cells, varying in form and function, but all share the same basic internal organization as those in the simplest animals and even plants. Within each cell is a distinct division of labor. Different cell processes occur within different types of compartments (organelles). Plant and animal organelles are similar with the exception of the chloroplasts, which animals lack. Thus, unlike plants, animals are unable to photosynthesize and make their own food. Plant cells also contain a rigid, cellulose cell wall, defining a shape but limiting the mobility and responsiveness that typifies animal life. All cells have a flexible cell membrane that holds in the gelatinous cytoplasm containing the organelles: nucleus, ribosomes, endoplasmic reticulum, vesicles, Golgi apparatus, mitochondria, vacuoles, and chloroplasts. In this activity, you will use clear gelatin and vegetables to build an edible model of a cell.

Cell Salad
(continued)

Materials
- heavy plastic bag (cell membrane)
- clear gelatin (cytoplasm)
- assorted vegetables (organelles): peas, carrots, jicama, beans, bell peppers, turnips, broccoli, mushrooms, celery, etc.
- clear plastic jar (cell wall for a plant cell)
- knife
- wavy chopper (optional)

Procedure
1. Decide if you are going to make a plant or animal cell.
2. Study the organelles necessary for your cell.
3. Mix up the clear gelatin, place in the plastic bag, and refrigerate.
4. If you are making a plant cell, place the plastic bag in a clear plastic jar.
5. Cut the vegetables to resemble the appropriate organelles.
6. As the gelatin begins to thicken, add the vegetables so that they are suspended in the gelatin.
7. Refrigerate until firm. While you are waiting for your gelatin to set, create a key for each cell part, showing the material used.
8. Your "cell salad" can be eaten or displayed for other students to view.

Middle School Science Challenge
© The Learning Works, Inc.

Cracker Wrapper

Force is any cause that changes the motion or the shape of an object. For example, when a person pushes the pedals of a bicycle, a force is applied and the bicycle moves forward. When someone squeezes a piece of paper to crumple it, a force is applied and the paper changes shape. These two examples demonstrate mechanical forces because two objects have touched one another. Gravitation is another force that acts upon all objects on Earth, pulling them toward the center of Earth. In this activity, you will be attempting to reduce such forces acting on a soda cracker by building an envelope or wrapper that will enable you to send a cracker through the mail and have it arrive totally intact and edible.

Cracker Wrapper
(continued)

Materials
- soda cracker
- one or two first-class postage stamps
- small scale (optional)
- building materials for the envelope/wrapper (possibilities include: envelope, paper, plastic wrap, bubble wrap, styrofoam cups, styrofoam pellets, strawberry baskets, wooden stirring sticks, straws, cotton batting, cardboard, scissors, tape, and glue)

Procedure
1. Wrap the soda cracker so that it will be edible.
2. Build a structure around the cracker to prevent it from being crushed in the mail. (The completed package must weigh two ounces or less.)
3. Place one or both of the stamps on the package.*
4. Address the package to your teacher using the school's address and mail it from your home.
5. When the package arrives, unwrap it with your teacher and assess the condition of the cracker.

*Note: An additional challenge may be to limit this package to one ounce (one postage stamp).

Middle School Science Challenge
© The Learning Works, Inc.

Discovering Decomposers

Animals that we know as herbivores eat only 10% of the plant material available. The other 90% is ultimately eaten by decomposers, which break down the organic material, returning the basic vital nutrients back to the soil and the ecosystems that rely on them. Without decomposers, nutrients would stay inside dead plants and animals and they, in turn, would continue to pile up all over the earth. Some of the organisms involved in decomposition include the following: bacteria, fungi, borer beetles and their larvae, flatworms, maggots, ants, termites, slugs, snails, worms, wood lice, millipedes, and mites. (Burrowing mammals assist by tilling the soil created.) The largest number of decomposers come from the bacteria and fungi groups. They feed by secreting digestive enzymes directly onto their food and absorbing the dissolved nutrients. The breakdown process is completed when they release simple compounds back into the environment so plants can use them again. In this activity, you will be examining a habitat and discovering decomposers.

Discovering Decomposers
(continued)

Materials
- habitat with plenty of plants (forest, riverbank)
- hand lens
- strainer or screen sifter
- paper
- pencil
- trowel
- pocketknife

Procedure
1. Find an area rich in plant life, preferably with some moisture or water nearby.
2. Begin looking for signs of decomposers, recording in words and drawings what you find.
3. You might scrape away part of a dead tree trunk or remove some of its bark with the pocketknife.
4. Dig down into the leaf litter and soil with the trowel.
5. Use the strainer or screen sifter to find organisms in the soil.
6. The hand lens can be used to help you find small organisms, trails left by them, or tunnels.
7. Look for fungi growing on logs, above ground, and under the leaf litter.
8. Look for the tiny threads extending beyond the fungi into the other plant matter.

Middle School Science Challenge
© The Learning Works, Inc.

Erosion at Work

Ancient Greeks and Romans wrote descriptions of erosion, noting the way rivers carry sand and clay to the ocean. Many ancient civilizations have cited the problem of erosion and its adverse effects on agriculture. Erosion is a constant process, forming new streams to carry water and sediments to the sea. In this activity, you will be able to model the passage of geological time, simulating the effects of thousands of years of erosion in a short period of time. You will be able to observe the effects of dams on rivers and to identify the following geological formations: waterfalls, riverbanks, riverbeds, tributaries, sediment, rapids, deltas, alluvial fans, levees, mouths, sources, landslides, meanders, lakes, forks, and valleys.

Erosion at Work
(continued)

Materials

- diatomaceous earth (pool supply stores)
- squirt bottle (like restaurant ketchup containers)
- plastic shoe box or wallpaper tray
- container for used water

- water
- sponge
- food coloring
- heavy cardboard

Procedure

1. Mix water with the diatomaceous earth in the shoe box or tray until it is soaked but not saturated.
2. Slope the surface. Use the sponge to remove excess water from the surface.
3. Fill the squirt bottle with water and a little of the food coloring.
4. Begin by dripping water from the squirt bottle, about one drop per second, at the top of the slope.
5. Continue for a few minutes, observing what occurs.
6. Try placing a piece of cardboard (to simulate a dam) across your stream.
7. Experiment with different slopes by propping up the fuller end of the tray.
8. Research the geological formations listed on page 76 and see how many you can identify.

Middle School Science Challenge
© The Learning Works, Inc.

Fossil Findings

Fossils are the remains of animals or plants that were preserved in rock before the beginning of recorded history. Several different types of preservation are known. One example is a mold. A mold is made from an indentation, such as a dinosaur footprint, in a sediment. A fossil remains when the sediment hardens into sedimentary rock. Molds of thin objects, such as leaves, are commonly called imprints. When a complete organism is found encapsulated and preserved in ice, like the now extinct woolly mammoth, it also is called a fossil. When minerals replace the cells of a buried organism one by one, a petrified stone fossil is produced. In this activity, you will create an archeological record of many layers of history, containing many types of fossils. You will then ask a classmate to examine and interpret your fossils.

Fossil Findings

(continued)

Materials

- plaster of Paris
- brown tempera paint
- petroleum jelly
- water
- sand
- one-quart yogurt container
- hammer
- screwdriver
- paper
- pencil
- large metal spoon or trowel
- materials for fossils (fern and other leaves, shells, bones, cardboard cut like feet, sticks)

Procedure

1. Gather your fossil materials and plan your layers, making a drawing as you go.
2. Coat all of the fossil materials and the container with petroleum jelly.
3. Mix the plaster of Paris with water and paint (to simulate mud) to a workable consistency. You will need to work quickly because the plaster hardens very fast.
4. Place a layer of plaster in the bottom of the container.
5. Place your first layer of fossil materials on top of the plaster with the most textured side down.
6. Cover completely with a layer of sand, which will aid in the separation of the layers later on.
7. Repeat with plaster, fossil materials, and sand until you reach the top.
8. When hardened, have another student analyze them, using the screwdriver and hammer to separate the layers.

Middle School Science Challenge
© The Learning Works, Inc.

Language of Life

Within each DNA molecule of a living organism there is enough information to carry on the development and maintenance of the organism. As with any language, DNA has a code that conveys information. The DNA language, or genetic code for making proteins, is actually quite simple, although each molecule is 13 feet long and tightly coiled within each cell. Each one of these strands of the DNA double helix is a chain of subunits or "words" held together like a ladder that has been twisted. The supports of the "ladder" are made up of sugar and phosphate units. The "rungs" or bases are adenine (**A**), thymine (**T**), guanine (**G**), and cytosine (**C**). **A** is always paired with **T** and **G** with **C**. Every DNA molecule is made up of many of these pairs in a unique combination or sequence. It is this sequencing that determines the "words" of the "language." In this activity, you will make a model of a portion of a DNA molecule.

Language of Life
(continued)

Materials
- Any items found around the house, garage, or yard (possibilities include ribbons, leather strips, fabric of any kind, shoelaces, pipe cleaners, wire, beads, washers, grommets, wooden dowels, embroidery hoops, gumdrops, toothpicks, and dried pasta)
Note: You will need to choose one material to represent each letter. Figure out how large a model you want to build and make sure you have enough of each item.

Procedure
Note: Use illustrated reference books to help you add detail and accuracy to your model.

1. Make a stand, or figure out a way to hang the double helix.
2. The double helix should be constructed so that it looks like a spiral staircase or ladder. Attach the sides of the ladder to the support, parallel to each other but twisted.
3. Add the rungs of the ladder in any order that you wish. Each rung will be half **A** and half **T** or half **G** and half **C**. It does not matter which side of the rung a particular letter is on.
4. Make a key to show what material you have chosen for each part of the molecule.

Middle School Science Challenge
© The Learning Works, Inc.

Liquid Layers

Why is it easier to float in the ocean than in a swimming pool? The answer is that salt water is denser than fresh water so it is better able to hold a person up. There is more mass in a given volume of salt water than there is in fresh water. If you took a cup of water and added two teaspoons of salt to the water, the salt would dissolve but the cup would now contain a greater mass than before the salt was added. Density is defined as mass divided by volume. Density allows us to compare two substances that have equal volume (meaning they occupy the same amount of space), but have unequal masses (meaning they contain different amounts of matter). It's like the old trick question, "Which is heavier, a ton of bricks or a ton of feathers?" You will never be caught by that one again because you know that the mass is equal but the *densities* of bricks and feathers are very different. In this activity, you will create a many-layered jar of household liquids based on their varying densities.

Liquid Layers
(continued)

Materials

- tall, narrow jar or glass
- small jar or glass
- food coloring
- household liquids (nontoxic)
- spoon
- salt
- paper
- pencil

Procedure

1. Choose a liquid and put a little of it in the jar. (You may wish to add food coloring to the liquid if the liquid is clear and water-based.)
2. Tip the jar slightly and slowly pour another liquid down the side of the jar. The denser liquid will end up at the bottom.
3. Record your results and empty the jar.
4. Experiment with different liquids until you have a list of relative densities. (Remember that salt can be dissolved in water-based liquids to increase density and that water and oil will not mix.)
5. Empty the jar once again. Begin layering your jar, starting with the densest liquid. Continue adding layers, working from the densest liquid to the least dense.
6. When you have finished, you will have an interesting-looking jar filled with bands of different colored liquids of different densities.

Middle School Science Challenge
© The Learning Works, Inc.

Map Making

Topographical maps show the shapes and features of the earth's surface. In order to show a particular elevation, a contour line is drawn across all of the land at that elevation. A conically-shaped mountain is shown by concentric circles. The distance between circles is determined by the elevation change at that point on the mountain, usually in intervals of one hundred meters. If you walk along a contour line, you stay at the same elevation, never going uphill or downhill. Sometimes there is an "X" as a benchmark showing a specific elevation, such as the peak of a mountain. Other features, such as rivers, lakes, roads, trails, forests, swamps, and buildings, are often shown as well. In this activity, you will make a topographical map of a clay "landform." As a class project, try to match each map to the correct landform.

Map Making
(continued)

Materials
- plastic shoe box
- modeling clay
- tape
- ruler
- pitcher of water
- overhead transparency
- permanent marker (fine tip)

Procedure
1. Set the box flat on the table. Place the ruler perpendicular to the table along the side of the box.
2. Mark lines, with numbers, every half centimeter, starting with zero at the table level.
3. Shape an interesting "landform" out of the clay and place it in the bottom of the box.
4. Using the tape, hinge the transparency along one long edge of the open portion of the box.
5. Looking straight down onto the "landform" and holding the transparency taut, trace the base of the landform onto the transparency.
6. Add water up to the half-centimeter mark. Trace the line where the water meets the "landform."
7. Continue adding water (in half-centimeter increments) and tracing the water line until the water covers the landform.
8. Remove your map from the box.

Middle School Science Challenge
© The Learning Works, Inc.

Rube Goldberg Machines

Rube Goldberg (1883–1970) was a well-known cartoonist who drew odd-ball inventions. *Webster's Dictionary* defines a Rube Goldberg as "a device or method to accomplish by extremely complex and roundabout means a job that actually could be done simply." These compound machines are made up primarily of several simple machines (e.g., inclined plane, wedge, lever, pulley, wheel and axle, and gear). In this activity, you will create and build a Rube Goldberg machine by putting together many simple machines. Possibilities include a page turner, an egg cracker, a plant-watering machine, a pet feeder, a golf ball washer, or a gumball dispenser. Of course, these are only a few of the unlimited contraptions you can build. Let your imagination run wild!

Rube Goldberg Machines
(continued)

Materials

There are two approaches to finding the best materials for your project. You can gather materials first, then figure out a machine that will utilize these items. The other approach is to design your machine on paper, then find the necessary materials. Good places to look for materials include the garage, the kitchen, desk drawers, craft stores, and hardware stores.

Procedure

1. Choose one of the two methods described above.
2. If you decide to design your machine on paper before you build it, use your sketches to make a list of the components you will need. After you have assembled your machine, make a final drawing of it showing only the components that you used.
3. If you build the machine first, make a drawing of your completed project showing the components that you used.
4. On your finished drawing, label all of the simple machines that you incorporated into your compound machine.

Middle School Science Challenge
© The Learning Works, Inc.

Sea Floor Soundings

The ocean floor is a relatively unexplored frontier. Most of our knowledge comes from soundings, which are depth measurements taken at many points. From these soundings, it is possible to make a picture of the ocean floor. The device used by scientists to achieve these soundings is called an echo sounder, or *sonar*. It produces a sharp clicking sound that travels to the ocean bottom and back at the rate of about 1,460 meters per second. By measuring the time needed for the sound wave to return to the ship, scientists can find the water's depth. For instance, if it takes two seconds to return to the ship, the water must be 1,460 meters deep. It takes one second to travel to the ocean bottom and one second to echo back. In this activity, you are going to simulate these soundings to map an unknown model of a portion of the ocean floor.

Sea Floor Soundings
(continued)

Materials
- plastic or cardboard shoe box
- modeling clay
- aluminum foil
- permanent marker (fairly fine tip)
- bamboo skewer
- graph paper
- ruler
- set of 12 colored pencils, crayons, or markers

Procedure
1. Set the box flat on the table. Using the modeling clay, form a sea floor with various elevations but none any higher than six and a half centimeters.
2. Completely wrap the box with aluminum foil. Exchange boxes with another student.
3. Measure the distance from the inside bottom of the box to the top foil. This is the deepest depth.
4. Make a key with each depth corresponding to a color in rainbow order, with violet as the deepest and red as the shallowest.
5. Write a zero in one corner of your graph paper. Number the squares along the short side and the long side. Use this as a template to mark the foil on the top of the box.
6. Begin making "soundings" by carefully poking a hole with the skewer until you strike clay.
7. Take a measurement to determine how much of the skewer has gone in and color the corresponding square for that depth on the graph paper. Repeat for all the squares.
8. Remove the foil and see how closely you were able to map the model sea floor.

Middle School Science Challenge
© The Learning Works, Inc.

Seed Success

Plants have evolved many ingenious ways of protecting and dispersing seeds, ensuring the future of the species. A fruit is formed around a seed as a protective case. Flowering plants form seeds in the female reproductive parts of the flower. Evergreen shrubs and trees, or conifers, bear and disperse seeds directly from the scales of their cones. Animals can disperse seeds by ingesting fruits and depositing the seeds in their feces in another location. They also can carry seeds with tiny hooks, such as burrs, that become stuck to their fur. Wind dispersal is common with seeds that have developed "wings" or are very light. Coconuts and alders are able to float across water and then germinate on faraway shores. When some plants become dry enough, they can actually pop out their own seeds. In this activity, you will be collecting a large variety of seeds, examining their means of dispersal, and planting them to see if they can germinate.

Seed Success
(continued)

Materials
- index cards
- tape
- water
- potting soil
- styrofoam cups
- pencil
- permanent marker (fairly fine tip)
- an assortment of collected seeds

Procedure
1. Go outside and collect as many different types of seeds as you can find.
2. Tape each seed to a card and record as much information as you can about its identity, the place it was found, and your guess as to its most likely means of dispersal.
3. For each seed, poke a hole in the bottom of a cup.
4. Fill each cup with moistened soil.
5. One at a time, remove each seed from its card and plant each seed about half an inch under the surface of the soil. Tape each cup to the corner of its corresponding card.
6. As each seed germinates and begins to emerge from the soil, record the date on the card.

Middle School Science Challenge
© The Learning Works, Inc.

Watching Indicator Wonders

An *indicator* is a substance that changes color according to the presence of a particular chemical substance or ion. There are several indicators that scientists use to show the presence of an acid or base in a liquid. In this activity, you will make an acid/base indicator and explore household liquids to see if they are acids, bases, or neutral substances.

Materials
- red cabbage
- knife
- clear plastic or glass cup
- straws cut in half
- water
- pan
- household liquids (nontoxic)

Procedure
1. Cut up red cabbage into pieces. Put the cabbage in the pan and cover it with water.
2. Boil the cabbage until a deep purple liquid is achieved.
3. Discard the solid matter.
4. Pour some of the purple liquid indicator into the cup.
5. Using a straw as a dropper, add one of the household liquids to the indicator in the cup, noting any color changes. It will turn pink with an acid, green with a base, and remain purple with something neutral. Try this with many liquids and various amounts of those liquids.

Field and Classroom
Experiments

Adapted to Dirt

Earthworms are in soil everywhere, helping to decompose dead plant material to make soil so that new plants can flourish. The dead plants that an earthworm eats are ground up in the worm's gizzard and passed through the digestive tube, emerging as castings. These castings enrich the soil with nutrients. The worm's blood is pumped through one main vessel from the head end to the tail end along the underside of its body. The blood returns to the head along the top of the body. In this activity you will observe how an earthworm responds to temperature changes in its environment by observing this top vessel.

Materials
- worms from the garden or bait shop
- a means to heat water
- ice
- aged tap water (24 hours)
- thermometer
- two clear, wide-mouth, plastic cups

Procedure
1. Place a worm in one cup, just covering the worm with cool water, no lower than 40°F (4°C). If necessary, place the second cup into the first to hold the worm in place.
2. Observe the speed and nature of the pulsing blood.
3. Continue to make similar observations, raising the water temperature each time. Do not go above 90°F (3°C).
4. Return the unharmed worm to the garden.

Solution to Adapted to Dirt

Results
You should have seen an increase in pulse rate as the water was warmed.

Explanation
Earthworms are poikilotherms. That means that they cannot regulate their body temperature. Therefore, their level of activity correlates directly with the temperature of the environment. You probably also noticed the worm's aortic arches, which are the closest thing(s) that an earthworm has to a heart(s).

Middle School Science Challenge
© The Learning Works, Inc.

Camera Connections

Cameras project images onto film and allow just enough light through to expose the film. How is the image of a large object focused on such a small piece of film? Which type of lens has the greatest curvature: a telephoto lens or a wide-angle lens? This experiment will help you learn the answers.

Materials
- plastic bag
- newspaper
- water
- two plastic hand lenses of different curvatures
- lamp with light bulb exposed (colored light bulbs work best)
- white paper

Procedure
1. Place the newspaper in the plastic bag.
2. Drop water on the plastic bag, one drop at a time, noticing how the curvature affects the magnification of the print.
3. Darken the room and turn on the lamp.
4. Hold one lens in front of the white paper, pulling it back until a focused image of the light bulb is formed on the paper.
5. Repeat this with the other lens.
6. Experiment with the light at different distances from the white paper.
7. Look at images around the room with each lens.

Solution to Camera Connections

Results

On the newspaper, you probably found that the rounder the drop, the greater the magnification. When you held the lens far from an object, you should have seen that the object was upside-down, as is usually the case with a camera. The lens with the greater curvature should have focused closer to the paper but created a smaller image than with the flatter lens.

Explanation

The lens makes something appear bigger because it bends the light, making us think that the light has come from a larger object. The more curved a lens is, the shorter the focal length. Thus, wide-angle lenses are more curved than telephoto lenses, which have a very long focal length but make large images on the film. In this experiment the white paper acted as the film.

Middle School Science Challenge
© The Learning Works, Inc.

Color Perception

Why do grapes appear green and cherries red? Why is it that in very dim light you do not see the colors of these objects but only their dark shapes? In this experiment you will determine why objects have a particular color.

Materials

- red cellophane (report folders work well)
- colored pencils, markers, or crayons
- paper punch
- tape
- index card
- two pieces of white paper

Procedure

1. Cut a one-inch square from the cellophane.
2. Punch one hole in the index card.
3. Tape the cellophane to the card so that it covers the hole.
4. Make a drawing of any object, using all the colors you have.
5. Look at this drawing and objects in the room with and without the cellophane filter.
6. Based on the information you just obtained, use all of your colored markers or pencils to write a sentence that can only be deciphered by looking through the filter. When looking at your message without the filter, each letter should be well hidden from the naked eye.

Solution to Color Perception

Results

Through the filter, the blues, greens, and purples should have looked dark and been seen easily. The reds, oranges, and golds should have been very light or not visible. Hopefully, you found that messages can be written with great secrecy by writing the letters in odd-shaped pieces of the colors that appear dark with the filter, filling in other "puzzle pieces" around the letters with mostly reds, oranges, and golds. Some bits of the other colors help to make the words less obvious.

Explanation

In 1672, Sir Isaac Newton theorized that light is made up of many colors. We observe this when a prism separates light into the colors of the rainbow. He also realized that in order to see the color of an object, the object must be illuminated. The object absorbs all colors of light except for those that are reflected to our eyes. Therefore, red cherries absorb all colors of light except for red, which is reflected from the cherries to our eyes. The red filter allows only red light to pass through to our eyes. Anything that is red reflects the red light, which passes right through the filter. Therefore, it appears bright.

Cricket Capers

Scientists classify animals according to their anatomy. Crickets are in the arthropod phylum and the insect class within that phylum. In this experiment you will observe live crickets to discover their anatomical and behavioral characteristics.

Materials
- cricket (pet store)
- clear plastic cup
- rubber band
- piece of apple
- hand lens
- netting
- stick

Procedure
1. Place the cricket, stick, and apple in the cup.
2. Immediately cover the cup with the netting and secure it with the rubber band.
3. Observe and record all of the physical and behavioral characteristics of the cricket that you can, using the hand lens if necessary.
4. Release the cricket into a field.

Solution to Cricket Capers

Results and Explanation

You should have been able to observe the following characteristics: The cricket has a hard, striped, waxy exoskeleton. It has three main body parts. On the head are two large compound eyes and three small simple eyes in between. The mouth is also located on the head. Hopefully, you were able to see the cricket eat the apple with its chewing mouthparts. Crickets have long, curved, antennae on their heads, made up of many small segments. The legs and wings are found on the thorax. A cricket has a total of six legs, all of which can be used for walking. The two hind legs, used for jumping, are much larger. The males can rub these legs against their bodies to make a chirping sound, which increases in speed as the air temperature increases. Behind the legs is the abdomen. If your cricket is female, it has a long straight ovipositor extending straight back. This is used to deposit eggs in the soil.

Middle School Science Challenge
© The Learning Works, Inc.

Determining Decay

Since many rocks in the earth's crust contain radioactive elements, scientists can use the measurement of radioactive decay as an accurate way of finding the absolute age of rocks. How are they able to do this?

Materials
- 100 pennies
- container with lid (e.g., one-quart yogurt container)

Procedure
1. Put all of the pennies in the container and cover it with the lid.
2. Shake the container.
3. Remove all of the pennies with the tails up. These have "decayed."
4. Record the number of "decayed" pennies.
5. Repeat this process as many times as it takes for all of the pennies to be used.
6. Determine what percentage "decayed" after each trial or "half-life."
7. Assuming that this container is a model of a rock for which you are trying to determine the absolute age and the penny's "half-life" is 10 minutes, how old is the container?

Solution to Determining Decay

Results

You should have found that about fifty percent of the pennies "decayed" after each trial or "half-life." To find the age of the container, you need to multiply the number of trials or "half-lives" that you made by 10 minutes. Therefore, the container would be about one hour old.

Explanation

Some elements have radioactive isotopes that are unstable, breaking apart or decaying at a constant rate. Half-life is the time it takes for half the mass of a radioactive isotope to decay. By comparing the percentage of a radioactive isotope with the percentage of the decay product, scientists are able to date rocks containing the original radioactive isotope.

Middle School Science Challenge
© The Learning Works, Inc.

Falling Freely

The ancient Greek philosopher Aristotle concluded that heavy objects fall faster than light ones. Can you make this same conclusion?

Materials
- two ping pong balls or two old tennis balls
- pointed scissors or knife
- lead shot, lead weights, or sand
- duct tape
- two pieces of notebook paper

Procedure
1. Carefully cut a slit in one of the balls. Fill it with the lead or sand, and tape it closed. Leave the other ball empty.
2. Place an equal-sized piece of tape on the empty ball so that both balls look the same.
3. Hold a ball in each hand as high as you can.
4. Drop both at the same time. Do they hit the ground at the same time or different times?
5. Repeat this several times to be sure.
6. Crumple one piece of paper, keeping the second one flat.
7. Repeat the experiment using the two pieces of paper. Are the results the same? Why?

Solution to Falling Freely

Results

Were you surprised that both the heavy ball and the light ball hit the ground at the same time, but the crumpled piece of paper hit before the flat one?

Explanation

Aristotle's hypothesis was commonly accepted until the late 1500s when the Italian scientist Galileo proposed a new idea regarding gravity. According to his idea, all objects fall with the same acceleration (change of speed) unless air resistance or some other force slows them down. It is air resistance that accounts for the difference in the time that the two pieces of paper hit the ground. The flat piece, especially when held in a plane parallel to the floor, encounters a great deal of resistance from the molecules of air in its path.

Middle School Science Challenge
© The Learning Works, Inc.

Picking Out Pigments

Most leaves are green, wouldn't you agree? Why, then, do we see yellow or red leaves in the fall? You might enjoy doing this experiment in the fall when the greatest variety of colors is available.

Materials

- glass jar
- pencil
- coffee filters
- tape
- acetone
- plant leaves (several different shades of green and other colors)
- spoon
- paper

Procedure

1. Use the spoon to grind one green leaf in the bottom of the jar.
2. Cover the ground leaf with about one inch of acetone.
3. Stir this for about five minutes.
4. Cut a strip of coffee filter paper as long as the height of the jar.
5. Place the pencil across the top of the jar.
6. Tape one end of the paper to the pencil, allowing the other end to drop into the acetone.
7. Leave the paper until the acetone has reached the top.
8. Lay the strip on a piece of paper.
9. Repeat with the other leaves.

Solution to Picking Out Pigments

Results
The paper strips from the green leaves should have bands of green across them. In addition, they may have bands of other colors (most likely, yellow). The papers for the yellow or red leaves would have no green bands, only bands of the leaf color.

Explanation
Chlorophyll is the pigment which gives green leaves their color and is responsible for photosynthesis. Often, the chlorophyll masks other pigments in the leaf, primarily the yellow carotenoid. When the leaf is no longer photosynthesizing in the fall, the chlorophyll breaks down, allowing the other pigments to be seen. Sometimes the red pigment is not formed until the chlorophyll has already broken down. The process that you used to separate these pigments is called paper chromatography.

Middle School Science Challenge
© The Learning Works, Inc.

Raging Rays

We have all heard about ultraviolet and infra-red rays from the sun striking the earth and causing harm. The greenhouse effect is one such concern. But do you know which kind of rays are actually responsible for this global warming?

Materials
- top or bottom of a fruit box (supermarket)
- white poster board
- table placed in the sun
- scissors
- tape
- thermometer
- large prism

Procedure
1. Turn the box on its side and set it on the table in the sun.
2. Cut the poster board in half lengthwise. Tape the two pieces together at the short ends.
3. Place this cardboard strip (your projection screen) inside the box to form an arc, perpendicular to the table top, with the ends at the opening of the box. Tape it in place.
4. Place the prism on its end between the sun and the projection screen until a rainbow appears in the middle of the screen.
5. Measure the temperature just beyond both ends of the visible light spectrum you see on the screen. Take several readings, making sure that everything remains positioned properly.

Solution to Raging Rays

Results
You should have found a higher temperature in the area beyond the red end of the rainbow.

Explanation
The sun's energy is radiated in all directions in the form of photons, or packets of solar energy. These photons travel in waves. We perceive photons with relatively short wavelengths as visible light. That is the light that the prism is breaking up into its component wavelengths, otherwise known as the colors of the rainbow. Red has the longest wavelength and violet the shortest. Other photons that have relatively long wavelengths and are perceived as heat when they strike us are the infra-red photons. The ultraviolet photons have an even shorter wavelength than the violet of the visible spectrum. In our atmosphere, water vapor, carbon dioxide, methane, chlorofluorocarbons (CFCs), nitrous oxide, and ozone are all considered to be "greenhouse gases" because they are transparent to visible light, but absorb infra-red photons, warming the atmosphere.

Middle School Science Challenge
© The Learning Works, Inc.

Sensitive Skin

A person's sense of touch is not consistent on all parts of his or her body. This experiment will demonstrate which areas of the body are the most sensitive to touch.

Materials

- two round toothpicks
- scissors
- index card
- modeling clay
- ruler

Procedure

1. Cut the index card into a piece that measures nine centimeters by four centimeters.
2. Using the ruler, draw a line down the middle of the length of the card and one across the card one centimeter from one end. Along the long line, starting with zero where the lines cross, mark and label each half centimeter up to seven centimeters.
3. Form the clay into a rectangle somewhat bigger than the card, about two centimeters thick, and imbed the card into the clay against one edge of the clay rectangle. Poke toothpicks through the clay at zero and at seven centimeters at the edge of the card, making sure the toothpicks are parallel.
4. Have someone else gently poke your back. Do you feel one or two toothpick points?
5. Keep moving the seven-centimeter toothpick closer to the zero toothpick until you feel only one toothpick point. Record the body part and the distance when this occurs.
6. Repeat this on various parts of the hands, arms, and legs.
7. Discard this instrument so that it will not be used on anyone else's body.

Solution to Sensitive Skin

Results
You probably found a great variation in sensitivity. The lips, fingertips, and palms of the hands are quite sensitive. The back and legs are not. Other parts are somewhere in between.

Explanation
Different areas of skin have varying numbers of nerve endings. Areas such as the fingertips must have many nerve endings because they are involved in sending many important messages to the brain having to do with touch.

Middle School Science Challenge
© The Learning Works, Inc.

Shaping Bubbles

We've all experienced and enjoyed soap bubbles—from the bubble wand to the bubble bath. What is a soap bubble, really? Technically, it is a thin skin of soap and water molecules encasing a gas, usually air or carbon dioxide from our breath. In this activity you will experiment with bubbles and think about why a bubble forms a particular shape.

Materials
- pipe cleaner
- bubble solution (or make your own by adding one cup of dish detergent to one gallon of water)
  Note: Some dish detergents are better suited to this activity than others.
  Experiment with different brands until you find one that works well.

Procedure
1. Bend your pipe cleaner to create a bubble wand with an interesting shape.
2. Make sure that the end is twisted back onto the handle.
3. Dip the wand into the solution.
4. Blow a bubble and observe the shape of the bubble.
5. Take your pipe cleaner apart and try the same thing with several other shapes.

Solution to Shaping Bubbles

Results and Explanation

All of the bubbles are spherical. A bubble forms a spherical shape because of surface tension. Water molecules have a positive charge on one end and a negative charge on the other end. Since opposites attract, water molecules align themselves so that they form a weak bond or attachment to each other. The soap simply acts as a means to reduce this surface tension by nudging between the water molecules, allowing them to stretch apart. The bubble will stop expanding as soon as a balance or equilibrium is reached by the air pressure inside and outside of the bubble's film. A sphere is the shape which allows the molecules to enclose the greatest volume within the least surface area.

Middle School Science Challenge
© The Learning Works, Inc.

Spin, Spin, Spin

As each planet revolves around the sun, it also rotates, or spins, on its axis. How fast is the earth spinning? How long does it take for the earth to revolve once? Why do we have different time zones? How many time zones are there? How wide is each time zone? What is the circumference of the earth? Is the earth spinning clockwise or counterclockwise, looking south from the North Pole? This activity will demonstrate the speed and direction of the earth's revolution and help you answer these questions.

Materials
- sharpened pencil
- four rocks
- large sheet of construction or butcher paper
- globe

Procedure
Note: Only two hours are necessary for this activity, although the results are more dramatic the longer it is done, especially when started early in the day.

1. Place the paper in a flat, open area that will remain sunny all day. Anchor the corners with the rocks. Stick the pencil into the ground in the center of the paper.
2. Each hour, trace the shadow of the pencil onto the paper and note the time beside it.
3. When you are finished, figure out where the shadows would fall for each additional hour. Use a ruler to draw dotted lines to represent these hours.

Along with a globe, you should now be able to answer all of the questions posed above.

Solution to Spin, Spin, Spin

Results and Explanation

The earth rotates once in 24 hours. There are 24 time zones. Each time zone compensates for the position of the sun relative to that area of the earth. Each time zone is approximately 1,000 miles across. The circumference of the earth at the equator is 24,900 miles. The earth, therefore, is spinning at a rate of more than 1,000 miles per hour! You should see that your tracings create one piece of a 24-piece pie each hour. Your tracings should also show you, by the movement of the shadows, that the earth spins in a counterclockwise direction as you look south from the North Pole.

Middle School Science Challenge
© The Learning Works, Inc.

Spin, Spin, Spin Confirmed

Do the stars move in the sky? Does the earth move relative to the stars?
You will need to do this activity on a clear, starry night.

Materials
- compass

Procedure
1. Facing south, find a place to stand so that you get a clear view of the sky but still have a reference point on the ground, such as a tall tree.
2. Choose a constellation that is easily recognizable and draw a quick sketch of its relationship to your reference point.
3. Repeat this each hour for two more hours.

Results
You should find that your constellation is moving across the sky from east to west.

Explanation
The earth is moving relative to the stars because it is rotating on its axis in a counterclockwise direction as you look south from the North Pole. That is why the stars only appear to move across the sky.

Discussion Starters

What is science?

This sounds like an easy question, but there must be a thousand answers to it, none of them fully satisfying. Some are correct but broad: "Science is the search for knowledge." Some are realistic but not helpful: "Science is what scientists do." Some are idealistic but misleading: "Science is the search for truth through the logical sequence of observation, formation of hypotheses, testing by experiment, and so forth, to devise general principles about the workings of the natural world." The problem is that none of these answers convey the highly diversified range of activities and styles of approach that science can involve. Field biologists and astronomers observe and record things. Chemists and physicists conduct experiments. Cosmologists may just sit and think. Would it help to define what science is not? What do you think science is?

Why are so many plant species becoming extinct, and how can we prevent further extinction?

One quarter (60,000) of the earth's flowering plant species are threatened with extinction. A large number of these grow in the tropics. Many of these plant species are invaluable sources of food, oxygen, beauty aids, fibers, resins, medicinal cures for diseases, and biochemicals for industry. Here are some of the suggestions that scientists have made to the countries where these tropical rain forests exist:

- set aside national parks to protect endangered species
- encourage selective harvesting and limit clear-cutting
- plant crops such as fruit and nut trees that provide food for many years
- protect the native cultures that have been able to live in harmony with nature
- encourage further scientific research to learn more about the area's ecology
- plant native tree crops such as coffee and cacao that do not harm the soil
- learn how to reforest areas that have been destroyed
- regulate slash-and-burn farming and cattle ranching

Is it possible to encourage natural preservation *and* meet the needs of humans? Analyze these suggestions and do research to determine what you think are the best solutions to this problem.

Middle School Science Challenge
© The Learning Works, Inc.

What is the solution to the problem of acid rain?

Acid rain is rain with a pH below 5.6 (a normal pH is 5.6 to 6.0.). Acid rain can kill fish in rivers and lakes, erode stone buildings and monuments, and destroy agriculture. Acid rain damages plants, soil, and water. As a result, the living organisms that depend on these resources are also put in jeopardy. Ninety-five percent of the acids in rain are sulfuric or nitric. Sulfuric acid is formed in areas where coal is burned for electricity. Nitric acid comes from internal combustion engines, especially automobiles. Some suggestions to solve the problem address the symptoms while others address the sources. Some ideas have been to build taller smoke stacks, add buffers to lakes, develop acid-resistant fish, increase the use of coal that is naturally lower in sulfur, remove sulfur and nitrogen compounds from fuels and emissions, substitute other energy alternatives for fossil fuels, and conserve energy. Research this subject further and come up with some of your own suggestions for solving the problems posed by acid rain.

What steps can be taken to prevent wildlife extinction?

Less than two hundred years ago, roughly one animal species became extinct each year. In the last one hundred years, the rate has accelerated so that we are losing several animal species *each day*. The rate of extinction is directly related to the rising human population. People need space for homes, farms, and commercial properties. Land is cleared to provide the needed space. Natural resources, such as timber, are harvested to support this development. In the process, habitats are destroyed, removing the elements vital for the survival of wildlife. Humans compound the problem by creating pollution, by introducing nonindigenous species, and through illegal and excessive wildlife trading. However, not all of the problems are as clear as those already stated. There is a strong correlation between the priority to protect animals and the standard of living in a particular country. Many of the world's countries are economically disadvantaged and thus have different priorities and interests that create additional pressure to use the available resources for survival. How might you approach this global problem to achieve a workable solution?

Is progress always positive?

Technology, the practical use of all areas of scientific knowledge, has exploded since World War II. Before this time, most technological advances were on a large scale, such as the development of the internal combustion engine and the nuclear reactor. Much of the newer technology is computer-based, involving complex processes and utilizing increasingly minute components. The silicon chip and microelectronics typify this trend. Technology has enabled doctors and researchers to improve their knowledge of diseases and how to fight them. At the same time, however, great debate surrounds the moral and ethical dilemmas presented by genetic engineering and prolonging patients' lives through the use of extraordinary means. Often, people don't stop to ponder the impact and implications of technology. Do you think that the advances technology provides are all positive or are there social, ethical, and environmental consequences with which people should be concerned?

What can be done about the hole in the ozone layer?

The ozone layer, which filters out the sun's harmful ultraviolet rays, exists in our upper atmosphere. In 1985, British scientists discovered a huge hole in this ozone layer over Antarctica. It is not certain what caused this hole, but many scientists believe that chlorofluorocarbons (CFCs) have eaten away at the upper atmosphere. These chemicals have been used in refrigerators and aerosol cans since 1941. Many countries have taken steps to discourage or eliminate the use of CFCs. Do you think that world government agencies should ban the use of CFCs entirely? Should more research be done to determine the cause of the hole in the ozone layer? Locate articles in the library or on the Internet to learn about the latest scientific thinking on this topic. Based on the findings of your research, what approaches do you think hold the most promise for preserving the ozone layer?

Middle School Science Challenge
© The Learning Works, Inc.

Should students with HIV be allowed to participate in school sports?

In recent years, the media has brought our attention to issues surrounding HIV/AIDS and sports. Basketball star Magic Johnson, former Olympic diver Greg Louganis, and other athletes have publicly discussed their HIV infection. This attention has led to an increased awareness of the need for AIDS education for athletes, but has also caused many people to develop fear about the possibility of contracting HIV during sports activities. Experts agree that there is a theoretical risk of HIV transmission from an HIV-infected player to an noninfected player if there is bloody contact during athletic practice or competition. However, studies released by the Centers for Disease Control in February of 1995 indicate that the risk is "extremely low." In spite of the extremely low risk, the Occupational Safety and Health Administration (OSHA) and other sports and health organizations have published universal precautions for handling blood from a sports-related injury. Do you believe that students with HIV have the right to participate in school sports? Should student athletes with HIV be required to disclose their health condition? Does the risk of infecting others (however slight) outweigh the benefits of athletic participation? Hold a classroom debate or a round-table discussion in which you address these questions.

What causes ice ages?

In the past 2.5 million years, there have been as many as 20 ice ages during which 40% of the earth's surface experienced a 100,000-year winter. However, the global temperature only drops 7 to 21°F. The breaks between ice ages typically last about 10,000 years, which means that the next one could be on its way. Several theories exist to explain this phenomenon, but none can be proven. Here are three of the theories that have been proposed:

- Orbital changes alter the seasonal amounts of solar energy received by each latitude. If this were true, summers would be cool enough to maintain ice and snow and the reflection off the ice and snow would reduce the temperature even further.
- Huge volcanic eruptions release so much dust into the atmosphere that the sunlight is blocked from reaching the earth.
- The earth's natural radiation output fluctuates because of changes in the temperature of the earth's core.

How would you assess these theories? Can you think of other possible causes?

How can we protect the environment from oil spills?

Crude oil and the products derived from it are vital to our way of life. The benefits of crude oil come with a price, however. On March 24, 1989, the supertanker *Exxon Valdez* ran aground off the Alaskan coast, causing a spill of 11 million gallons of crude oil that created an ecological disaster area along hundreds of miles of Alaska's coastline. This event brought much publicity to some difficult questions. Should supertankers be allowed to transport crude oil in our seas? Should the construction of the ships be regulated? Should such potentially harmful shipping be regulated more carefully? When an oil spill occurs, who should take responsibility for the destruction to nature? What do you think we should do to protect our environment from such disasters in the future?

Answer Key

Page 10
a. Jupiter
b. icy particles of different chemicals
c. methane gas in the clouds
d. 0.001%

Page 11
a. chemical reactions with oxygen inside living cells
b. about 100 trillion
c. spiders' silk
d. sea horse, *Hippocampus breviceps*

Page 12
a. ring system
b. Antarctica
c. two-thirds
d. 136.4°F in Libya (in 1922)

Page 13
a. hexagonal
b. gold, copper, silver, tin, and lead
c. In most cases, the light is passed through a ruby.
d. It doesn't melt; it sublimates or evaporates.

Page 14
a. bacteria (3.5 billion years ago)
b. light
c. four
d. spider

Page 15
a. liquid
b. sodium
c. Water pressure increases with depth.
d. The volume of an object can be measured by the amount of water it displaces.

Middle School Social Studies Challenge
© The Learning Works, Inc.

Answer Key
(continued)

Page 16
a. around the North and South Poles; frozen
b. one-quarter
c. the Gulf Stream; the Gulf of Mexico
d. Krakatoa

Page 17
a. electric eel
b. ant
c. cheetah
d. whale

Page 18
a. the hottest, where complete combustion takes place
b. At high altitudes, water boils at a lower temperature.
c. from a plane or high cliff
d. aluminum

Page 19
a. blonde
b. 19–34 days
c. lungfish
d. They have heat-sensitive cells on their heads.

Page 20
a. varies from one to three times the size of Earth
b. by studying a cylinder of ice, almost a mile and a half long, drawn from the Antarctic ice cap
c. It would be hotter and 20% drier.
d. zircon (about 3.8 billion years old)

Answer Key
(continued)

Page 21
a. Capsaicin, a chemical that makes food spicy, stimulates nerve endings directly, producing "reflex" perspiration.
b. tooth enamel
c. nitroglycerin
d. REM—an abbreviation for rapid eye movement

Page 22
a. the inert gas, neon
b. in the center of a thermo-nuclear fusion bomb
c. a laser beam striking a sapphire crystal
d. atmospheric or barometric pressure

Page 23
a. among the coral reefs off northwest Australia
b. larva of the polyphemus moth, *Antheraea polyphemus*
c. large tropical scarab beetles
d. egg cell from the female ovary

Page 24
a. Doppler effect
b. flexible bamboo poles used as levers
c. radon
d. gold

Page 25
a. aurora borealis, northern lights
b. another star's gravitational attraction (It will take *Voyager 2* at least 27,000 years to travel far enough for that to occur.)
c. Jupiter
d. Tahiti

Middle School Social Studies Challenge
© The Learning Works, Inc.

Answer Key
(continued)

Page 26
a. brain
b. hummingbird
c. feet
d. male monkey

Page 27
a. carbonated water
b. by flying at a higher
   altitude
c. liquid helium II
d. Einstein's Theory of
   Relativity

Page 28
a. Bay of Fundy in Nova
   Scotia, Canada
b. tsunami
c. Tutunendo, Colombia
d. gneisses (about four billion
   years old)

Page 30
a. Marie Curie
b. Sir Isaac Newton
c. Dmitri Mendeleev
d. James Joule

Page 31
a. Rosalind Franklin
b. Gerty and Carl Cori
c. Louise Brown
d. Hippocrates

Page 32
a. Edwin Hubble
b. Benjamin Banneker
c. James Hutton
d. James Hansen

Answer Key
(continued)

Page 33
a. Count Amadeo Avogadro
b. James Watt
c. Orville and Wilbur Wright
d. Ellen Ochoa

Page 34
a. Archimedes
b. Max Planck
c. Dorothy M. C. Hodgkin
d. Albert Einstein

Page 35
a. Antonia Novello
b. Alexander Fleming
c. George Washington Carver
d. Rachel Carson

Page 36
a. Mae C. Jemison
b. Ptolemy
c. Nicolaus Copernicus
d. Johannes Kepler

Page 37
a. Jonas Salk
b. Thomas Hunt Morgan
c. Francisco Bravo
d. Louis Pasteur

Page 38
a. Samuel Morse
b. Franklin Chang-Diaz
c. Theodore Maiman
d. Gabriel Fahrenheit

131

Answer Key

(continued)

Page 39
a. Valentina V. Tereshkova Nikolaeva
b. Neil Armstrong and Edwin Aldrin
c. Benjamin Franklin
d. Guion S. Bluford, Jr.

Page 40
a. Georg Ohm
b. William Thomson, Lord Kelvin
c. Steve Jobs and Stephen Wozniak
d. Amelia Earhart

Page 41
a. Luc Montagnier and Robert Gallo
b. ichthyologist
c. Charles Drew
d. Leonardo da Vinci

Page 42
a. George Gamow and Ralph Alpher
b. Jocelyn Bell and Anthony Hewish
c. Galileo Galilei
d. Maria Mitchell; Mitchell's Comet

Page 43
a. Theodore Lawless
b. William Harvey
c. chimpanzees
d. Elizabeth Britton

Page 44
a. James Clerk Maxwell
b. Lise Meitner
c. Granville Woods
d. Irene Joliot-Curie

Answer Key

Page 45
a. Anna Botsford Comstock
b. Charles Best and Frederick Banting
c. Daniel Williams
d. Mary Leaky

Page 46
a. Mary Mallon, nicknamed Typhoid Mary
b. Ernest Just
c. Dixy Lee Ray
d. Ivan Pavlov

Page 47
a. Florence Van Straten
b. Alessandro Volta
c. Jack Kilby
d. Antoine Lavoisier

Page 48
a. Andreas Vesalius
b. Charles Darwin
c. Margaret Sanger
d. Robert Hooke

Page 50
a. circulatory system
b. EEG, electroencephalogram
c. hemoglobin
d. circadian rhythm

Page 51
a. calorie
b. center of gravity
c. proton
d. flash point

Middle School Social Studies Challenge
© The Learning Works, Inc.

Answer Key
(continued)

Page 52
a. Moh's scale
b. Van Allen belt
c. trenches
d. metamorphic

Page 53
a. ovipositor
b. epiphyte
c. echinoderms
d. stipe

Page 54
a. troposphere
b. oxygen
c. halo
d. continental shelf

Page 55
a. bit
b. solid
c. radio waves
d. chemical bond

Page 56
a. lichens
b. sponges
c. reflex
d. colon

Page 57
a. lever
b. oxide
c. hydrometer
d. angle of incidence

Answer Key
(continued)

Page 58
a. eclipse
b. quartz
c. petroleum
d. Precambrian

Page 59
a. Arthropoda
b. pseudopodium
c. producer
d. alveoli

Page 60
a. alloy
b. quasars
c. surface tension
d. potential energy

Page 61
a. the Ring of Fire
b. estuary
c. meteors
d. moraine

Page 62
a. biosphere
b. genetic engineering
c. amino acids
d. phloem

Page 63
a. condensation point
b. nuclear fission
c. atomic number
d. electrolysis

Middle School Social Studies Challenge
© The Learning Works, Inc.

Answer Key
(continued)

Page 64
a. basal metabolism
b. viruses
c. chromosomes
d. stamen

Page 65
a. minerals
b. continental divide
c. jet streams
d. tectonic plates

Page 66
a. decibel
b. inertia
c. chemical equilibrium
d. ionic bond

Page 67
a. ozone layer
b. frontal zone
c. mantle
d. Pangaea

Page 68
a. cambium
b. Hominidae
c. chlorophyll
d. nucleotides